MARS

Heather Hammonds

CONTENTS

Rigby

THE PLANET MARS

Sometimes, the planet Mars can be seen in the sky at night. Mars looks like a bright star.

Mars looks like a bright star.

But when we look at Mars through a big telescope, we can see that it is a planet!

moon

Mars

Sun

moons

Earth

Mars is the fourth planet
from our sun.
It is smaller than Earth.
It has two little moons.

north pole

ice cap

ice cap

south pole

Mars has seasons,
just like Earth.

Mars has a north **pole**
and a south pole, too.
We can see
white **ice caps**
on the poles.

A Cold and Stormy World

Mars looks like the dusty deserts on Earth, but it is much colder.

We cannot breathe the air on Mars. It has a different kind of air to Earth.

Canyons and craters
on Mars.

Mars has **canyons**,
mountains, and **craters**.
It also has big clouds,
dust storms, and fog.

are giant volcanoes on Mars.

A volcano on Mars.

Millions of years ago,
the volcanoes on Mars were **active**.
Now they are **extinct**.

Mauna Loa

Olympus Mons

Olympus Mons is the biggest volcano on Mars.
Mauna Loa is the biggest volcano on Earth.
Look how big Olympus Mons is!

10

DRY RIVERS

Small **spacecraft** have been sent to Mars.

The spacecraft have taken pictures of dry rivers.

Some of the dry rivers have islands in them.

Dry rivers on Mars.

Scientists think that Mars
once had a lot of water.
But now it is a very dry planet.

PEOPLE ON MARS

One day, **astronauts** may travel to Mars. It will take them many months to get there.

They will have to wear spacesuits
when they go outside their spacecraft.
But, for now, only machines visit Mars.

GLOSSARY

active doing something. Active volcanoes are hot, and have lava and ash flowing from them.

astronauts people who go to space

canyons deep valleys

craters big holes made by meteorites

extinct died out. Extinct volcanoes are cold. They don't have lava and ash flowing from them.

ice caps ice that sits on top of the north and south poles of Mars

pole the top or bottom part of a planet

spacecraft a machine that travels through space

INDEX